WORMS

Nature's Formations

How Are Islands Formed?

B.J. Best

Cavendish Square
New York

Published in 2018 by Cavendish Square Publishing, LLC
243 5th Avenue, Suite 136, New York, NY 10016

First Edition

CPSIA Compliance Information: Batch #CS17CSQ

All websites were available and accurate when this book was sent to press.

Library of Congress Cataloging-in-Publication Data

Cataloging-in-Publication Data
Names: Best, B.J.
Title: How are islands formed? / B.J. Best.
Description: New York : Cavendish Square Publishing, 2018. | Series: Nature's formations | Includes index.
Identifiers: ISBN 9781502625373 (pbk.) | ISBN 9781502625397 (library bound) | ISBN 9781502625380 (6 pack) |
ISBN 9781502625403 (ebook)
Subjects: LCSH: Islands--Juvenile literature.
Classification: LCC GB471.B47 2018 | DDC 551.42--dc23

Editorial Director: David McNamara
Copy Editor: Nathan Heidelberger
Associate Art Director: Amy Greenan
Designer: Alan Sliwinski
Production Coordinator: Karol Szymczuk
Photo Research: J8 Media

The photographs in this book are used by permission and through the courtesy of: Cover Oreo Nuras/EyeEm/Getty Images; p. 5 Thomas Nord/Shutterstock.com; p. 7 Pikappa51/Shutterstock.com; p. 9 Daulon/Shutterstock.com; p. 11 Planet Observer/UIG/Getty Images; p. 13 G. Brad Lewis/Aurora/Getty Images; p. 15 Chris Clor/Blend Images/Getty Images; p. 17 Nil Kulp/Shutterstock.com; p. 19 Roland Bouvier/Alamy Stock Photo; p. 21 Jhorrocks/E+/Getty Images.

Printed in the United States of America

Contents

How Islands Are Formed **4**

New Words **22**

Index **23**

About the Author............ **24**

An island is made of land.

Water **surrounds** it.

Islands can be in the ocean.

They can be in lakes.

Earth is made of **plates**.

Plates are huge pieces
of land.

That land can be under
the sea.

The plates fit like a puzzle.

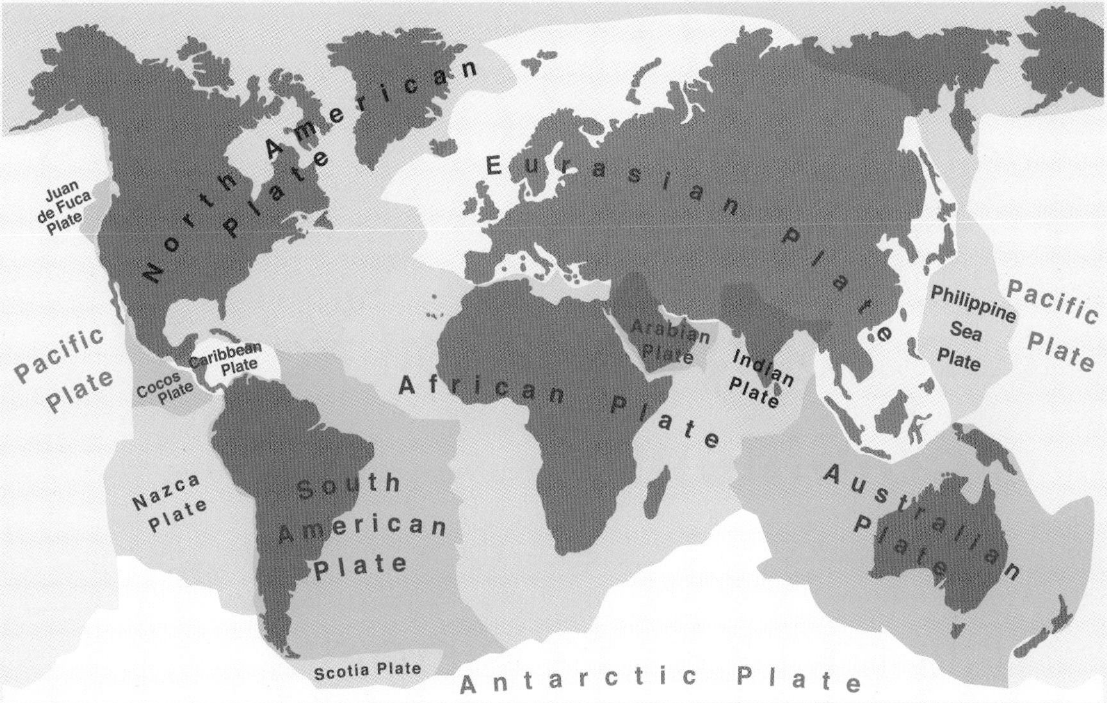
North American Plate
Eurasian Plate
Juan de Fuca Plate
Pacific Plate
Philippine Sea Plate
Pacific Plate
Arabian Plate
Indian Plate
Caribbean Plate
Cocos Plate
African Plate
Nazca Plate
South American Plate
Australian Plate
Scotia Plate
Antarctic Plate

The plates move very,
very slowly.

Land can break off from
a **continent**.

This makes an island.

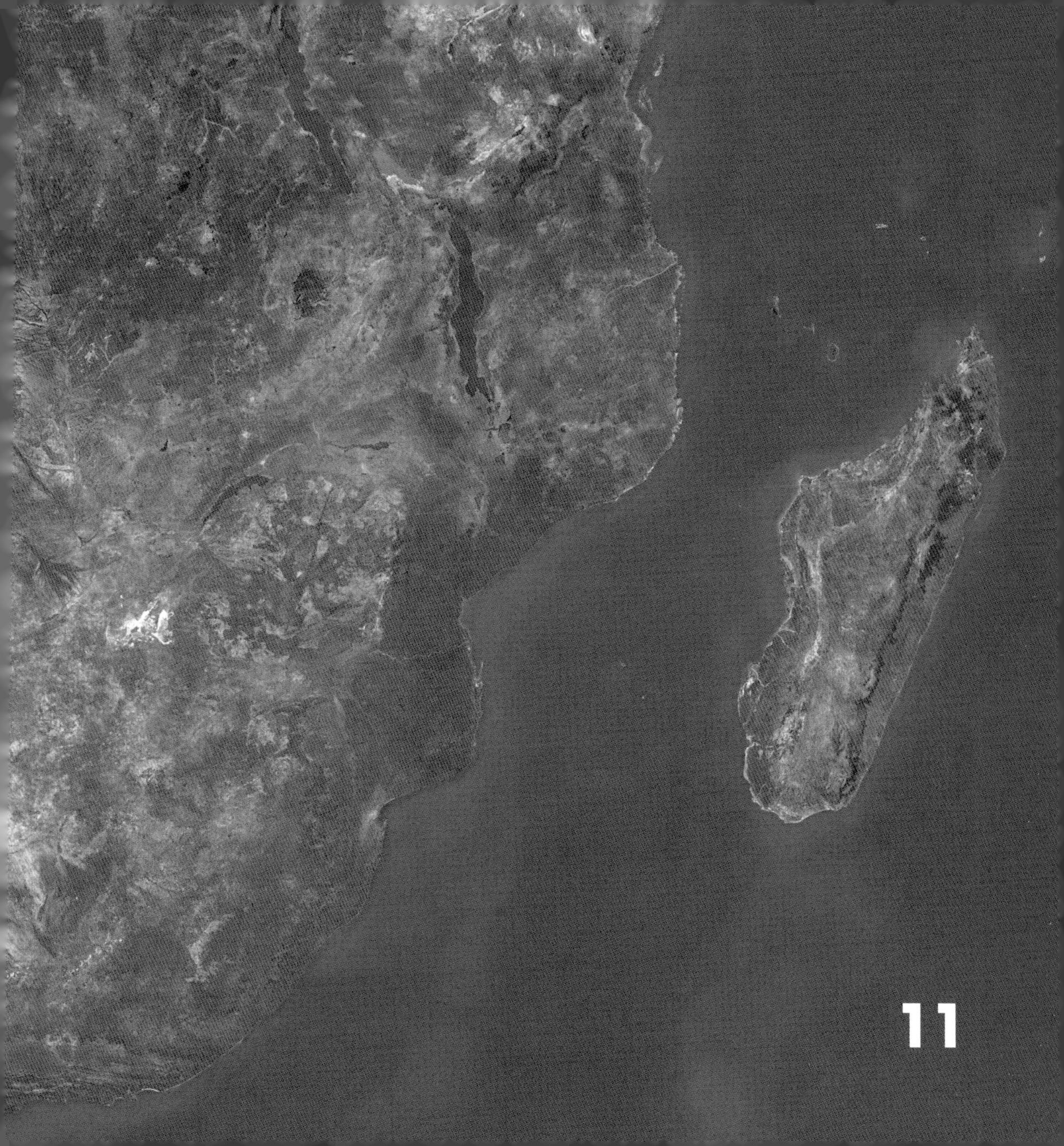

Molten rock is under the plates.

Molten rock can come from below the sea.

It builds up. It cools to become solid.

The new rock gets above the water.

An island is formed!

The amount of water can change in lakes.

High water can turn hills to islands.

It takes a very long time to make an island.

Today, people live on and visit islands.

They like to be near water!

New Words

continent (CON-tin-ent) A huge land mass.

molten (MOLE-ten) Melted.

plates (PLAYTS) Huge pieces of land and sea.

surrounds (suhr-ROUNDZ) Is on all sides.

Index

continent, 10

lakes, 6, 18

molten, 12, 14

plates, 8, 10, 12

rock, 12, 14, 16

surrounds, 4

About the Author

B.J. Best lives in Wisconsin with his wife and son. He has written many other books for children. He lives near a lake that has four islands.

About BOOKWORMS

Bookworms help independent readers gain reading confidence through high-frequency words, simple sentences, and strong picture/text support. Each book explores a concept that helps children relate what they read to the world they live in.